edZOOcation™ presents
Gorilla

by Sara Karnoscak

Early Pre Reader

Gorillas have thumbs.
They help them grab.

You have thumbs, too!

Gorillas are strong.
They can lift things.

You are strong, too!

Gorillas are smart. They learn lots of things.

You are smart, too!

Gorillas play tag.
They like to have fun.

You play tag, too!

Gorillas can laugh.

They laugh when they get tickled.

You laugh, too!

Gorillas eat fruit.
It keeps them healthy.

You eat fruit, too!

Gorillas can roar.
It makes them sound big.

You can roar, too!

Gorillas can feel shy. They may want to be alone.

You can feel shy, too!

Gorillas are not monkeys.

They don't have tails.

You don't have a tail, either!

Gorillas sleep at night.

20

You sleep at night, too!

Sleep tight!

23

Dedication:

For Cody, strong and gentle.

−S.K.

For Oliver, playful and smart.

−T.S.

Copyright © 2023 Wildlife Tree, LLC. All rights reserved.

Author: Sara Karnoscak

Designer: Tiffany Swicegood

Editor: Tess Riley

With special thanks to Meri Lawrence, zookeeper, for her expert review of this book.

Photo Credits:

AdobeStock.com

Pixabay.com

Pexels.com

ISBN: 979-8-9859544-5-6

This book meets **Common Core** and **Next Generation Science Standards.**